RED PILL

With no flags and no faces, behind the cyberattacks of the financial world

ORLANDO DÍAZ

ISBN: 978-1-7359917-0-2 (eBook Spanish edition)
ISBN: 978-1-7359917-1-9 (Paperback Spanish edition)
ISBN: 978-1-7359917-4-0 (eBook English edition)
ISBN: 978-1-7359917-5-7 (Paperback English edition)

CONTENT

CONTENTS

1

DOWN THE RABBIT HOLE

On May 24th, 2018, one of the main banks in Chile was the victim of the largest cyber-attack to date in the country.

Thousands of workstations and hundreds of servers failed, the malicious attacker program prevented the computer recovery by damaging the *Master Boot Record* [MBR], thus erasing the first sector of the computer's hard drive.

A domino effect occurred, and it affected, first, tens, then thousands, until reaching around ten thousand computers.

Completely black screens, unavailable systems, annoying clients everywhere, and this was just the beginning of the nightmare. An infamous attack with the *MBRkiller* program had taken place.

What people at the bank did not know while undergoing this nightmare was that this action was only a play, surgically prepared, to make a masterstroke, and conclude with the theft of ten million dollars.

The attackers used the chaos to target the *SWIFT* transfer system, extracting money from the bank by manipulating the records in ways that made them untraceable.

This was not their first attack, though. They had already done it before, between 2015 and 2018, with the SWIFT system. The SWIFT, which stands for the «Society for Worldwide Interbank Financial Telecommunications», interconnects more than eleven thousand financial institutions in more than two hundred countries, and through it, trillions of dollars in money orders travel daily, was in the crosshairs of hackers.

RED PILL

Months before the attack in Chile, Bancomext in Mexico had been given a close warning in Latin America, when a group of hackers tried to steal the not so small figure of one hundred million dollars. It was disguised as a small "donation" from the bank to a Korean church; however, the operation was unsuccessful. In addition to identifying the anomaly in a timely manner, luck was on their Mexicans' side, since the receiving bank on the other side of the world was closed due to the early hours of the morning, this allowed the transfer orders to be stopped and reversed.

These sophisticated and silent threats carefully executed by groups known as APT (Advance Persistent Threat) take the necessary time to advance within the network without raising suspicions until the criminal can achieve their objective.

Some of the attacks on the Latin American financial system originated after the posting of malicious code on the website of the Polish banking supervisory authority. The malicious code was also found in the entities of the National Banking and Securities Commission in Mexico, as well as in a state bank in Uruguay. The objective? Performing a "watering hole" attack in the system.

Obviously, having chosen the supervisory entity of the Polish Financial System was a smart move, since the institution had generated a lot of interest about the legal implementation, or not, of cryptocurrencies in the country, and other banks from all over the world were keeping an eye on this situation.

The carefully planned "watering hole" attack did not affect all the institutions it was intended to, but it managed to affect financial institutions in Mexico, Chile, Brazil and Colombia, and then it focused on banks and to a lesser extent on Internet and telecommunications companies. After this "discrimination" process, nine objectives in Mexico and six in Chile had been selected.

The malicious program used in the attack took advantage through JavaScript, to exploit weaknesses in Microsoft Silverlight and Flash Player while users browsed the affected web pages, subsequently proceeding to download the backdoor stored on the legitimate sites that were already equally compromised.

Attacks by these groups use a variety of techniques to compromise banks, ranging from spear phishing to waterhole attacks, prioritizing highly stealthy attacks against their targets.

Typically, in these phishing attacks, they send well-crafted deceptive emails in which they include all types of details, from financial information to job offers that require visiting a malicious web link or downloading and opening a dangerous document. But it is not the typical phishing email message that we are used to, full of typos or foolish requests. In recent cases, it has been found that they even take advantage of the LinkedIn platform to make their attacks more convincing regarding job offers.

These malicious links exploit *0-day* weaknesses in the system, and they can go unnoticed by security solutions, with the aggravating circumstance of protecting the code with commercially available tools such as the «*Enigma Protector*».

After establishing itself, the malicious program proceeds to escalate privileges until it reaches its final objective, taking advantage of network segmentation flaws.

To stay undetected, it uses in many cases the same tools available from the operating system, such as *SC* and *NETSH*, and in this way establishes its backdoor in the memory to avoid leaving traces of unknown files on the hard drive. In case of requiring administrative privileges, they usually take advantage of the credentials of existing services and reuse them with tools such as *Mimikatz*.

Its permanence in the network is protected by masking connections to command and control *(C&C)* and imitating secure connections on the web via *TLS*, taking advantage of an attack known as *«Fake TLS»*. Which makes detection very difficult for any security system, in addition, communicates with multiple proxies that generate a new *TLS* connection until reaching its destination.

Of all the attacks directed to the financial system, the most striking has undoubtedly been the one that occurred in 2016 against a bank in Bangladesh, with which the hackers managed to escape with no less than around eighty million dollars, a good blow without a doubt, but nothing compared to the nearly one billion dollars they tried to transfer.

A blow from which they were "luckily" unable to extract the rest of the money because one of the indications in the transfer coincided with the name of an oil company that was on Iran's sanctions list, which caused that thirty of the thirty-five transactions were canceled and sent to revision for further scrutiny.

A very well-elaborated attack that penetrated the institution's most sensitive system, where the hackers were able to place a malicious "malware" program to control and manipulate the bank's software on the SWIFT network, allowing them to use the messaging tools as desired, as well as manipulating the evidence and transfer records that were constantly printed throughout the day.

It has been one of the biggest financial robberies in history, with an impressive technical level.

These attacks are neither common nor simple, they require surgical preparation, penetrating to the heart of the bank and evading security systems until finding the right moment to act. Once the perfect moment arrives, a chain of international transfers then begins, which can end up in casinos in the Philippines or in other countries with poor anti-money laundering standards, allowing the trail to disappear. And then comes the destruction indication, which can be either surgical or not, as long as it manages to remove all relevant records.

Detecting these attacks is extremely difficult, due to the use of various techniques of obfuscation, code rewriting, and application of commercial encryption tools to hide them.

Security protection solutions give us a false sense of security when dealing with attackers of this level. Banks and commercial companies do not, though they think they do, play in the same league as the authors of these threats. The most effective prevention mechanism is to assume that you will be robbed and that you must have your recovery plan ready.

The resources and capabilities of these attackers are not just around the corner, so the perpetrators are usually never found, but this does not detract from the euphoria and motivation to discover the techniques and tools used by the attackers, as well as being able to track them down.

Luckily and unintentionally, fate had somehow placed me on the map to support this hunt.

2

SILICON SONS

As it happens with any high-impact incident, the services of international companies were contracted and security agencies were involved to carry out a forensic audit.

The investigation, as in other cases, determined the complexity and elaboration of the incident to circumvent the security mechanisms and thus usurp the credentials of the bank's SWIFT system. It was found that it was not a direct attack on the SWIFT network, but on the financial institution and its protection mechanisms to use this service.

An old friend, named Robert, was in Chile the day of the attack and was invited to a conference of a manufacturer of security products, and what he least imagined was that his visit would turn out to be more interesting than expected.

A few days after the incident, a professional colleague from the prosecution or another Chilean institution, of which I did not receive many details, shared with Robert some general impressions about past experiences and similarities that could help connect the dots with the incident at the Banco de Chile.

"They used a malicious *0-day* program to destroy the workstations," said Santiago, his friend in Chile.

"It's never been used before, but the cues seem to indicate that it's a modified version of earlier attacks," he continued.

"They're skillful," Robert exclaimed, acknowledging the quality of the attackers.

"I wish you the best of luck," he said ironically.

"They altered and erased the audit logs, they basically became a ghost," Santiago said.

"Fortunately, during the outage caused to the computers, the system identified abnormal transactions in the SWIFT system so they were able to stop them, which helped prevent the financial blow from being even greater," said Santiago.

"Obtaining information from the code through reverse engineering is highly complicated, they used *VuProtect* for the coding, but there are some coincidences with a module of the malicious program named *Buhtrap* that could guide us," he concluded.

Robert frowned and looked at him, then he threw a light smiled and replied:

"Sooooo basically they don't have anything. Better make them an altar and their respective reverence to those sons of the silicone."

—"What the heck, dude, why praise them? They are causing us a lot of problems! I want you to examine the incident, unofficially, of course," said Santiago.

"I'd love to!" Robert responded excitedly.

"I'm sure you have more skills and knowledge than the professionals hired for such a task... They are good, don't get me wrong, but it is not the same to have years of professional training in the field and make you living from it than being a self-taught person breaking the rules and testing what the systems do beyond what they were designed for. Guys like you play at being gods on the net," Santiago finished.

Robert didn't deny it, but at the same time he knew that the height of the bar in these attacks was set too high, even for him, but this would only serve as gas to fuel his desire to get his hands on the evidence.

Despite his enthusiasm to be a part of history, he was doubtful and calculated if accepting this task was the right move. Doing this for fun or work with manageable risks would immediately place him on the stage, and in this situation, he was dealing with an organization that was not attacking for mere political or personal ends, but a powerful criminal organization.

He knows that confronting hacktivists is not the same as facing organized crime organizations or those supported by governments, and due to the level and advanced nature of the attack, we deducted that they were "lone wolves", hackers... They were not *script kiddies*.

Robert was aware that these types of organizations did not respect any limits, and the consequences of exposing himself might not only be suffered in the virtual world.

It is not common to see such advanced attacks to obtain economic gains, since these tactics are usually carefully protected to be used exclusively in acts of espionage to government or private companies, to steal military secrets or intellectual property. In a nutshell, situations resemble a lot a virtual cold war.

This discrepancy and unusual situation only fueled Robert's desire to get involved. To have the opportunity to closely observe his skills, to see his creativity and ingenuity and to be able to identify the attackers' tactics and procedures, and perhaps getting to the origin of the attack. Refusing the opportunity was something unimageable for him.

Understanding Robert's reasons requires that we know a little about his past.

3

FIRST JUMP

The Matrix was released in 1999, and it was the first time we saw a movie that dealt with hackers from a grounded perspective. Despite many interpreting the film as pure science fiction, martial arts, and entertainment focused on special effects, the movie dealt with the topic of artificial intelligence and the computer world way much better than previous hacker films.

Unlike other movies that feature exaggerated screens and messages that looked like the cover of a celebrity magazine, this one managed to amaze us with the subtle presentation of the *Nmap* tool, *the Network*

Swiss Army Knife, shown in the movie for as long as a blink. The movie also showed the use of an *exploit* simulating the violation of the SSH communication protocol, used by Trinity to deactivate the emergency system of the electrical network in the film. There are many other references to the world of computing, science and even biblical, but that's another story.

Details worth paying attention to speaking of how extraordinarily prepared the screenplay of the film was regarding the technologies, even though they refer to very tiny details that appear on the screen for just a couple of seconds. Understanding and enjoying those moments beyond the special effects were not an easy task for those who do not share the world of silicone.

Very few movies have been able to make a real interpretation of a computer attack. Most of them seem like designers were told to draw funny and confusing images for the big screen. Nobody said it would be easy, some had good attempts, like the Snowden movie, but most ended up having terrible interpretations like the ones in *Swordfish* and *Skyfall*. It took fifteen years for a series to respectably show what hacking really looks like, in *Mr. Robot* with his "*Fsociety*."

That year, our favorite band *Rage Against The Machine* spoke fire with their words at the WoodStock concert. His lyrics against injustice and the system achieved a unique connection between us, which would serve as a boost for our technological anarchy.

It's no coincidence that the first *The Matrix* movie ended with the *Rage's* song called "Wake Up," and we had no doubt that we were awake.

Robert and I, for several years, were developing our skills together and reinforcing our knowledge, both with groups of Black and White Hackers.

I met him while he was breaching the security of one of the systems I was protecting, instead of feeling anger or frustration, we felt a connection right away.

We went to college together to get bored, but we took advantage of our love of computers to entertain ourselves.

We dreamed about going to the *MIT* (*Massachusetts Institute of Technology*), Stanford, or Berkeley University in California; these universities were very advanced in terms of technology and were doing things that we wanted to get involved in, but our resources did not allow us to pay those high tuitions.

That did not prevent us from continuing to dream; the talents that these universities exported were incredible. Bill Joy at Berkeley, during his graduate school, created the *BSD (Berkeley Software Distribution) Unix*. He then founded Sun Microsystem, the rest, after billions of dollars later, is history.

Ken Thompson and Dennis Ritchie also studied at Berkeley, they were, nothing more and nothing less, than the fathers of Unix and the C language, basically the language that allows the operating systems we know today. Some engineers who are credited with great Xerox inventions and who brought to life the technology that surprised Apple, such as the mouse and the graphical interface, which it later copied, also studied there.

MIT also had some of its pupils to brag about, its classrooms had seen the development of Tim Lee, who took credit for being the father of the *World Wide Web*; William Bradford, for the invention of the transistor, and

Richard Stallman, founder of the *free software* movement and the *GNU* operating system.

And Stanford? Well, it was one of the four nodes where the *ARPANET* originated, and where the design of the Internet and the *TCP/IP* protocol began, and if giving life to what we know as the Internet is not enough, it was in this university where *Google* was born, from a research project conducted by Larry Page and Sergey Brin.

Basically, a large part of technology and most of the most important technological companies that we know were impacted in one way or another by talented people who passed through these universities, in which, without a doubt, their methods and tools allowed to increase exponentially the capacities of their students.

We hated the university, but we would definitely have wanted to study in these.

Well, we didn't really hate college, but the antiquated industrial-age educational system that was, and still is, implemented to educate students in most parts of the world.

Robert was a super interesting character and possibly a genius. We had fun in a different way, instead of having an afternoon of video games, we listened to the telephone conversations of everyone who was lucky enough to make a call near the radio cells of the cellular network of our locality, using a code of access on Motorola *StarTac*.

Doing it was extremely simple as long as you knew the code and the sequence of characters that you had to execute, but that information was not available to everyone, so we boasted saying a phrase widely used by all who shared our interest: «*Knowledge is power*».

Getting in and out of third-party systems was seen as a challenge and healthy competition and having a few remote servers around the world with backdoors modified to use them as your *hop points* or personal *proxies* turned into endless conversations.

Robert sometimes went a little further, we never did anything for profit, but he did enter our school system to improve his class grades, for example. In the end, I guess it was a more efficient way for him, instead of dealing with know-it-all teachers and their super boring classes where he could hardly learn anything. He was so self-taught and confrontational that, on one occasion, he had read the entire Bible in one weekend just to discuss an argument with a church father.

But he mostly used his knowledge for trivial benefits. I don't forget his personal version of Mario Bros, in which he had modified the mushroom from the video game to increase in size by a joint of cannabis, and when he obtained it, in addition to increasing in height, he grew Afro hair in the best style of Jimi Hendrix or Michael Jackson in his early days.

He could make these changes in a matter of minutes, open his favorite code editor, load the binary file, identify the value of interest and replace it in seconds. And he was able to do this while having a conversation and listening to loud music, as if his fingers worked independently of his brain without being interfered with by external factors.

His thirst for knowledge, as well as his arrogance were inseparable, and he always found a way to get me amazed with his occurrences.

On another occasion, at dawn on a new year, I went to wish him a happy new year and found him working on the computer, in a room so messy that the floor was barely visible. He had no idea what time it was, he wasted no time in celebrations, he was someone who exclusively focused on his priorities.

This obsessive dedication, love and passion for computers and technology is what differentiates these individuals from the rest, and the reason why it is almost impossible for professionals who have studied at the best universities and obtained any type of diploma, or the highest degree of qualification to be at the same level of these individuals.

That dedication wasn't healthy either unless it's the goal of your life. Our obsession left no room for socializing, so learning to dance, going to parties, drinking alcohol without reason or control like many young people, or enjoying some sport, they were very foreign things to our minds, so they used to call us antisocial or nerds.

But to be honest, it was never like one of those scenes from a movie in which people remained locked in a basement without seeing the sun. We had our friends who did not share our passion and helped us maintain the necessary balance in our lives and meet the classic nonsense of youth.

With them we could talk about different topics, but more than half of the time we felt that it was a waste of time, but little by little we began to value behaving simply like boys, making jokes, drinking alcohol until we laughed uncontrollably... I can say, to a certain degree, that we began to enjoy it.

The interaction with new people was not so simple, we preferred to analyze and study them, like a Freud with a laser sight. We hardly ever heard someone say something that got us interested in starting a conversation, so we didn't do very well with small talk.

Our technological journey was not a sea of roses either, we spent our days seeing our modems set on fire while confronting those with more knowledge who attacked us, and spending days configuring a new one was not so much fun. *"Plugging"* and *"playing"* were not friendly activities in the Linux system at the time.

While people were getting tribal tattoos on their skin, we got a tattoo of *Tux*, the *Linux* penguin, in honor of our love for our favorite operating system on the planet. We were victims of strange looks on the beach as a penguin tattoo was always noticed weird. We even had to dissuade a drunk man by making gestures to indicate that we were heterosexual.

The experiences, the more challenging they were, the more they provoked something in us, they fueled the fire of the desire to improve ourselves and led us to devour thousands of pages in a few days.

There was no malice in the crazy plans that Robert came up with, just an uncontrollable appetite for knowledge. He didn't act like a cracker, circumventing the security of systems and vandalizing companies or *websites* in search of recognition. His style was more like a ninja quenching his thirst for knowledge.

Robert sweated the hacker's manifesto: *"This is our world now, the world of the electron and the switch...my crime is curiosity...and being smarter than you..."*

Santiago, his Chilean friend, knew Robert's background and therefore was confident that he could help him more than anyone else he knew, and for this reason, he entrusted him with a copy of the affected hard drives and several critical logs of the attack on Bank of Chile.

Excited by the task, Robert devoted himself day and night to the research endeavor, his face reflected on the screen revealing the intensity of his gaze and his concentration, his fingers giving orders and instructions nimbly as if they had a life of their own. He barely ate well, and often forgot to shower, he could wear the same clothes for several days without giving it any importance, sometimes he slept only two hours and even put a potty chair under the desk so as not to waste time visiting the bathroom.

That's the kind of obsession you won't find in a professional who does the job for a paycheck. It is not that he is healthy either, but it is always better to do something you love...

After an intense marathon, he had not managed to discover anything else. He knew as much as the research group had already reported, he had reached the same conclusion, but it took him way less time. He had to break the bad news to Santiago.

Still, the hunt for him was far from over.

4

STAY FOOLISH

Around the year 2000, we met Khalil, who had already started in this field and had a great desire to continue learning, but in our opinion, he had a long way to go before we could pay attention to him. However, his attitude convinced us to welcome him to our group.

We all had to work together, so the interaction served to nurture us all, and he, in particular, acquired knowledge at a surprising speed.

Khalil was another *nerd*, and he was even more reserved and religious than we would like him to be. Gradually, he learned many of our secrets and became one of us, but without the ties of brotherhood that Robert and I had.

It was the same time as *Back Orifice 2000, "BO2K"*, developed by the *Cult of Death Cow* as an administrative tool, which was later exploited to perform some small crimes on third party computers. It allowed almost complete control of the victim through a "nice" graphical interface (GUI), in the sense that it was simple and easy to perform, since the interfaces at that time were extremely horrible.

But most of us got here after experiencing Linux consoles and their beautiful SHELL interface, with no GUI at all.

Linux users have always ridiculed Windows users who wanted to be hackers using that system.

This program, BO2K, allowed many non-computer-savvy people to amuse themselves by performing the pranks they saw in movies, such as opening or closing programs, controlling the keyboard and mouse, or spying on user activity on the computer. Khalil was one of those who had fun with BO2K and we saw him as power in the hands of the unworthy.

It wouldn't have been so bad if they at least ran a Light version of Linux on top of Windows, or set a *dual-boot*. It would have helped even with the fresh versions of *Ubuntu* and *Suse Linux*, which at that time were like the babies of the house. It wasn't that we were expecting them to run a full *Red Hat system*, either.

But it did not leave a good taste in the mouth to see the ease with which any individual, just by knowing how to move the mouse, could be making himself believe as the most *badass* hacker that had ever existed.

A year later we could no longer say the same about Khalil, at that time he had learned a lot and he even took some time to tutor us on specific topics.

We didn't call anyone a hacker, especially because we respected the origin of the word; the ingenuity, skill, creation and contribution of these individuals to the computing community. Individuals with a vision to build their own world, and anyone who wants to participate.

But then the news started calling computer vandals hackers, when they really meant crackers, and well, as they say, the rest is history. It's not easy to compete when you're fighting the "fourth estate" in a pointless fight.

By this time, Khalil was finding 0-Day weaknesses in operating systems and programming his own exploits without much trouble, so it was obvious the progress he had made.

Khalil had also intensified his criticism of the political system, and his tone was more anarchist than ever, showing no intention of sitting idly by.

Robert and I didn't fault the idea, but we weren't vandals, so we didn't add fuel to that fire, mainly because we thought it was enough of a deterrent to discourage him from taking radical action without us all being on the same page. We didn't like public exposure.

Or so we thought, until sometime later, on the way to work like any other day, something unusual happened.

I felt in the air of that Monday a strange tranquility, different from the chaotic traffic of my city.

Arriving at the office, the tranquility became tense, it was not a normal Monday.

The employees were secretive, theorizing, until I finally found out.

"You saw how crazy the message is, it's in all the news, including international ones," commented Raúl.

"What are you talking about?" I asked.

"What? Haven't you seen it? Come, walk," he replied in a rushed manner.

It was then that I understood the novelty of the day, the reason for that particular atmosphere.

A massive defacement had taken place, including the major news outlets and other major companies, the most important being the electoral system website.

By that time, most of the country knew about the attack and the messages transmitted by its authors.

The message against the judicial and political system generated a lot of interest. Something that many would have wanted to shout from the bottom of their lungs, due to the chaotic management of political management.

You could almost visualize the anger and fury on the face of whoever had written the message.

The authors concluded in a burlesque, arrogant and cynical tone, paraphrasing the hacker's manifesto and the superiority of those who carried out the attack.

I almost expected those quotes, but the last sentence left me cold: *"I am nothing, no one, nobody."*

5

CHOICE IS AN ILLUSION

That short phrase that expressed disidentification and pretended to hide in anonymity, like the mask of *Guy Fawkes* immortalized in the movie V for Vendetta, told Robert and me something else.

That phrase wasn't part of the hacker's manifesto, but it was from a *Rage Against the Machine* song called "Maria." The phrase referred to the "voice of the unheard," about illegal immigrants who come to America looking for a better life, who have to put up with the worst possible abuses in the jobs they get because of their irregular conditions.

"Maria" did not represent our favorite song, it did not compete with the strength, intensity, and lyrics of "*Killing in the name*", which deals with the issue of racism and police brutality. That song really shocked us, a song from 1992 that is still valid decades later. Another song with a deep message was "*Know your enemy*" that spit fire against the war, and the abuse and hypocrisy of the authorities.

However, "María" had that particular phrase that we used when we achieved our goals, supported by anonymity and the lack of propagandistic promotions. There was a sense of grandeur behind its use, the lack of identification was not a show of humility, and we just couldn't believe there was so much coincidence, which led us to think of Khalil.

We questioned Khalil without getting him to admit his actions. We spent hours relaxing and then aggressive, we fought and laughed, but he never admitted doing it. But this coincidence and his increasingly anarchistic tone didn't help us believe a word of it.

In the following days we could not continue our attention on this matter, because Robert must have felt the same cold that he had felt when reading the final sentence of the defacement, when he saw how his computer turned black, with the following message: «*Non-system disk or disk error*», without needing to read the rest he said «replace and strike any key when ready».

"Fuck!" he uttered. "What the hell happened here?" He added.

You do not have to be a genius to understand that he was living in his own flesh, on a personal scale, one of the effects of what had happened at Banco de Chile. The *MBRkiller* was back.

He was angry and confused, "*how the hell did he get through without triggering any alarms?*" he thought.

He did not underestimate the security of the bank, they have to deal with many network services and interconnections between computers that make protection work difficult, the balance between not obstructing the business and having good security controls is a constant struggle that often causes many security professionals to lose their hair. This is a problem that gets even the best chiefs of cybersecurity or *CISOs* on their nerves.

But in Robert's case, he didn't provide any services, and his computers and network were prepared to fend off any attack, or so he thought.

The most relevant difference in his audit, versus that carried out by the companies contracted for forensic analysis, was that Robert's computer where he ran the analysis was connected to the Internet, since he did not have to follow the strict protocol of those who were contracted.

Now, this little help towards the Chilean friend had become something personal.

It would no longer be a passive analysis; it would be more like Gary McKinnon's obsession with hacking hundreds of US military servers just to find out if they were hiding information on unidentified flying objects (*UFOs*).

Robert's anger was more related to the damage that had been caused to his ego than those materials. He did not save any important file on the affected computer and repairing it was a matter of little effort, however, that did not imply that he neglected security measures. He had to consider that this computer could serve as a bridge of infection or attack for the rest of his computers.

He spent days going over in his head everything he had done and what he had analyzed in the cloned operating system; memory, temporary data, hash values, encrypted files, etc.

He analyzed the *MBRkiller.nsi* and the *Buhtrap* malicious program code where a similarity had previously been identified, with the objective of identifying references or indications that could be useful to the investigation. He also explored the *VMprotect* code to identify weaknesses that could allow him to decrypt part of the protected code. For a moment he thought of exploring the encrypted part again, but this time he decided to change the way he approached the problem, looking for a new perspective.

Paranoid as he was, Robert not only had intrusion prevention systems (IPS), *firewalls*, and file integrity monitoring systems, but he also had an event handling system capable of reconstructing all network activity.

This helped him capture and store every single activity on the computer. It was like repeating a scene from a movie until understanding every detail.

It was at this moment when he decided to check the network traffic, right before the incident. A quick view did not bring up much interesting information, just the expected OS traffic and no unknown protocol connections. It was mostly *HTTP* and *HTTPS* traffic from their web browsing.

Something didn't add up to him, *"irregular traffic should be obfuscated in these protocols"*, he thought.

In the following days, almost without closing his eyes, he sat down to analyze the thousands of records produced during the hours prior to the attack, without being sure what parameter he was looking for. His ego just wouldn't let him sleep.

Exhausted, tired, and upset, he decided to involve me in his odyssey.

"Come to my house, I have something important to tell you," Robert said.

I didn't even ask him what it was. I knew that if it came from Robert, I could be pretty sure to be something interesting, so I ran to his house.

"Let's see, who did you crack today?" I asked.

"Well, it seems that it was the other way around " Robert exclaimed with laughter.

Then he started telling me everything in detail.

Robert was more technical than me, so when working together, what I could offer him was another strategy or point of view. Seeing the process in an untainted way, he could question everything as if it were a blank page, and I could offer him options that perhaps, due to his obsession, he had overlooked.

I spent hours with him checking the logs he showed me with his progress or lack thereof, it was very difficult to find an anomaly in traffic that appeared to be normal. I must admit it... Hackers have become very good at the art of deception.

We proceeded to change the approach and, instead of checking the communication traffic in a general way, we began to apply filters to eliminate what we understood could be discarded. This plan had its disadvantages, since if the malicious program had spoofed a file that we interpret as safe, we would be excluding it from the analysis, but we decided to try.

After filtering the communication from hundreds of records, the results were reduced to a few dozen lines related to Twitter.

It didn't seem interesting at all, until I asked him:

"Did you post or download an image during the forensic analysis?"

"Not that I remember," he replied.

Remember or not, an image had been downloaded via Twitter. At first, we thought it was a simple graphic file that took advantage of the use of stenography to hide information.

But further analysis helped us understand that this was a Polyglot, which allowed a compressed file to be disguised within an image, and with the additional encoding, made it extremely difficult to detect.

"They're good..." I exclaimed.

"Good sons of bitches," Robert replied.

While we were analyzing, we identified that they had used Twitter as their command-and-control source, obtaining their indications through post and hashtag readings to decode the malicious file, thus preventing it from being detected by security tools. And subsequently, in the same way, they get the command action to kill the boot system of the computer. They had run a variant of HammerToss to run the MBRkiller.

When we got the IP address where the malware had connected from, we shouted with excitement.

We were euphoric because we had identified a significant part of the attack, but on the other hand we knew that, despite our enthusiasm, hackers capable of carrying out attacks of this level did not live by giving away their final IP addresses simply because we had managed to understand how one of their little toys worked.

What awaited us now would be nothing compared to what we had experienced in the past.

6

WHAT IS CONTROL?

The world has already seen the effects of government-sponsored cyberattacks. In 2007, Estonia, one of the most connected countries in the world, was practically shut down.

The country, called *"The Silicon Valley of the Baltic"*, had a mobile penetration of 107% and 81% of homes connected to the Internet in 2007.

With its technological culture, «e-Estonia» has built paths for innovation and the use of technologies for the daily use of its population. At that time, more than 70% of people used banking services and 97% of transactions were carried out online. In a country with 1.3 million inhabitants, 1.1 million had online banking accounts. It was, and still is, an impressive number under any scenario.

We can just say that, as a reference, more than a decade later, first world countries are still struggling to reach these figures.

In 2002, Estonia issued their first "ID Cards" using digital signatures for their citizens, and by 2006 they had issued more than a million. This ID card is used for everything from transportation to online tax filing.

It was the first country to hold electronic voting during the 2005 national elections. They voted using computers!

Skype was created in Estonia before *eBay* bought it for 2.6 billion dollars in 2005, at the hands of the same developers of *Kazaa*, the *Spotify* of that time.

This small, highly connected, and techno-logical nation suffered a massive denial-of-service (*DDoS*) attack on its technology infrastructure for three weeks, reducing the capacity of government and the daily lives of its citizens in an unprecedented way.

In a nation so dependent on the Internet, not having it to provide services to its citizens was a catastrophe.

Almost all government ministries were attacked, institutions that highly related to the presidency, the media, telephone companies, banks, as well as any service that the attackers considered relevant.

It is estimated that more than a million computers were hijacked and turned into zombies to create a huge botnet used in this attack. A million computers, statistically, was almost the equivalent of one for every inhabitant in the country. The distributed denial of service (*DDoS*) attack flooded the Estonian servers to the point of drowning them.

It was not until a select group of individuals got involved, including a very particular one, Kurtis Lindqvist, responsible for managing one of the Internet's thirteen Root *DNS (Domain Name Servers)* when the fate of Estonia began to change. When the Internet traffic of the entire nation was about to collapse, Kurtis, who was on vacation in the country, began an action plan that allowed him to contain and gradually end the attack, thanks to his privileged access and knowledge.

This attack was accredited to Russia, as it coincided with Estonia removing a World War II memorial monument. Russians did not like that.

As we evaluated the level of sophistication of the techniques used in the attack on Robert's computer, we were quick to relate them to the capabilities of the attack on Estonia, even though the technique was not the same.

"Have we moved any sacred monuments?" Robert asked.

"I'm not sure, but everything indicates that we have run into a *Nation-State hacker*," I replied.

Robert made a face as if he had just smelled cow dung.

"You think so? It does not make any sense. They only attack governments or industries with sensitive information that can be useful to them, or place *backdoors* in critical infrastructure for when war calls. For what reason would they expose themselves by attacking banks? They don't need the money. It doesn't make sense," Robert replied.

"I don't know, but who the hell has this attack power, planning, and patience?" I replied.

Robert looked from side to side, lowered his head and immersed himself in his thoughts without answering the question.

After a while he said:

"I insist, it would be very abnormal, why would they do it? Russia? China? Iran? The United States? Look at the Chinese operation with *Shady Rat*, who spied on more than seventy companies around the world for at least five years. If it were up to them, they would go unnoticed for a hundred years. Exposing their techniques and how they use the vulnerabilities that the industry does not know about is not a smart decision, they don't have an infinite store of 0-days exploits either," Robert concluded.

Robert was referring to one of the largest intrusions ever made, which affected the Federal Governments of the United States, Canada, Vietnam, Japan, Switzerland, among others, as well as energy, IT, satellite, military security, politics, and economy companies, from all over the word, including technological giants with enviable security measures like *Google*.

It was considered one of the largest transfers of wealth in history that happened under everyone's noses, mainly affecting the United States, although it had a global impact. They stole national secrets and sensitive government data, as well as source code and intellectual property from corporate companies. In a nutshell, they took what we can imagine and even what we will never be able to imagine. Some say that part of China's technological advances came after this *"shortcut"* they took.

"I don't know," I answered again.

"But I don't know of criminal gangs with these levels either, maybe some *hacktivists* would endeavor in something like this, but it doesn't make sense either," I concluded.

The reason why the *Nation-State hackers* decided to attack a bank puzzled us, but we did not rule it out until we then identified a new powerful and dangerous gang of criminals.

The *Nation-State hackers* do not play in a different league than common *hackers*, they really play in another galaxy, as we can see in the attacks suffered by some countries. While Estonia had been digitally shut down, Ukraine was literally shut down with an attack on its energy grid.

That was his early holiday gift, on December 23, 2015, some cities in Ukraine met the *BlackEnergy* malware family, hundreds of thousands of Ukrainians experienced darkness for the first time. A malicious program had penetrated the network of the industrial system that made it possible to control the electrical energy switches, and from there manually send the instructions to cut the energy and interrupt the service for several hours.

Interestingly, at one of the power stations, the modus operandi was very different and even amusingly Machiavellian; using *Radmin*, the energy company's technical support tool, they blocked the keyboard and mouse actions of the operators, so that the attackers had total control over the switches and, thus, could turn off each station of the city.

Stunned operators watched as the mouse handled itself and proceeded to turn off each switch, one by one. Later, without being visible, the *hackers* destroyed backup systems and corrupted computers to hinder the emergency response. And to complete the chaos, they attacked the telephone lines with a bombardment of calls to prevent effective communication during the event.

The electric company's technical team had to transfer the substations from automatic to manual mode, thus closing the power system switches. Restoration of all services took between three and six hours.

This is something we could have seen it in a poorly written and directed movie, but sometimes reality is stranger than fiction.

In 2016, around the same date, there was another attack, another blackout, but in the capital of Ukraine, Kyiv. This time they did not attack distribution stations, but a 200-megawatt (MW) main transmission station, far exceeding all station attacks in 2015.

The system was out for an hour, affecting at least a fifth of the city, long enough for the pipes to begin to freeze and the situation went from being uncomfortable to a life-or-death condition.

This new improved version of the malware was able to attack and control *Siemens* devices in power stations and was created in such a way that it could be reused in other countries. It would be the second known malicious program that could go from the digital world to the physical world, the specialists affectionately nicknamed it *"Industroyer"*.

This ability to reuse and export was discovered in 2014, in the Department of Homeland Security of the United States. Previous versions of the malware had been identified in their critical infrastructure since 2011. Its presence has been found in different servers, including ones controlling oil, gas, electric power, and water distribution, as well as wind turbines and even nuclear plants. A powerful ace up the sleeve in case of war.

If there was any doubt about the power of a cyber-attack, these samples ended up clarifying the matter, and put the great powers on earth on alert.

This is the new policy of the world order, the new nuclear weapons.

In these attacks on Ukraine's power grid, planned around a year earlier, they used carefully designed spear *phishing* tactics. The first emails did not even contain malicious programs, but they did contain references to *PNG* image files within *HTML* codes, each image customized for its victim.

Did they just open the email, or did they load the entire message with their images? It was a very careful victim recognition strategy, and not the typical case of common attackers who tend to attack by taking advantage of the moment.

After understanding the email reading behavior of the employees, in the following months, hackers moved on to the second phase of the plan, sending a new *spear phishing* attack, this time focused on delivering the malicious program, a simple *Office* file; either in *Word*,

PowerPoint or *Excel*, containing a *0-day* exploit with the malicious instructions inside a macro code, thus turning a common day-to-day file into the source of infection.

After successfully exploiting the first computer, they continued with the reconnaissance phase, obtaining system information and network perimeters; version number of the programs, current privileges, processes, etc. In many cases, they took advantage of the same operating system tools used for administrative tasks.

They managed to obtain the password them from the system, browsers, email credentials, etc, through theft and data capture. From there, they began to move to other computers until they obtained administrative credentials on the network, and with all these gathered, they prepared the checkmate.

Companies spend large sums of money to protect themselves from cybersecurity threats, and sometimes, they just create obstruction and friction among their own services and technological innovations. In the end, the threats that materialize the most are those where the weakest link in the chain is exploited, the human factor.

In the attack carried out in Ukraine, due to the tactics and techniques used, it is technically very difficult to be able to identify the origin of the attacks.

In our case, when we obtained the *IP* address of the attack on Robert's computer, the geographical area of the attack concerned us a little, so we began our exploration of the target as stealthily as possible; for some reason it reminded us of the events in Estonia and Ukraine.

It didn't take us long to find weaknesses in the *web* server of such *IP* address and to be able to access the computer, which ended up being another pawn in the chain.

Detecting the *rootkit* installed on the computer to secure unauthorized access remotely took us a little longer. The activity in the logs were surgically removed to disguise the owner's records. So we set up our own *kernel-level* monitoring system so it wouldn't be fooled by the previously installed backdoor.

After several days we found that the backdoor was receiving connections from the Czech Republic and from China, but what caught our attention were a few from

Russia with relatively high TTLs (Time-to-live) responses from a specific IP address.

We continued our exploration to the new *IP* address, even more carefully, without expecting that it was the network address of the protagonists, even though we felt that we were very close.

And we received the confirmation.

7

FEELING THE ENEMY

A massive *DDoS* attack had been launched towards our network, the computers had just died, they weren't responding to anything.

The protection measures were useless, in the end, this was hardware competition, or rather, the lack thereof. We didn't have the capacity to handle that bandwidth, there was no way for us to compete.

The attack was not limited to a few minutes or hours, but to days, and as expected, it not only affected us, but all the lucky ones who shared our network block. We could not estimate how many thousands of innocents were affected as well.

The Internet provider was working like crazy to solve the situation, but it's difficult when the enemy has no head nor tail. We opted to acquire the service from another provider, because *Wi-Fi* first, and then food, were on our list of priorities, according to *Maslow's* new pyramid.

The attack was the final confirmation that we were at war, and it seems that we overlooked the level of the opponent, from now on nothing would be very paranoid. We would run from a *live USB* using the highly confidential operating system called *Tails*, based on the famous *Debian*, built specifically for privacy and security. Using *TOR* by default for all incoming and outgoing communication, not just the *Web*, and blocking all non-secure communication, and using *LUKS* for *USB encryption*, while operating entirely in memory without leaving any trace on the computer. It's the same system recommended by Edward Snowden, famous for publishing classified NSA documents, including his Internet surveillance programs *PRISM* and *XKeyscore*.

We had to sacrifice our comfort to advance in this war, using one of the most secure operating systems in the world, one not intended to be your daily operation tool, since it does not store any type of information and eliminates constantly the information it uses.

We would use multiple *hop points* in different countries to further protect our *IP addresses*, but we would also do it from public networks via *VPN (Virtual Private Network)*, in shopping malls or cafes, so that it would be virtually impossible to track us after each connection. Furthermore, we would add a couple of tablespoons of patience, or rather, tons, because all this would impact our performing speed, and we were not famous for being patient.

Now that we were in open war, we did not only reinforce the defense, but also proceeded to reinforce the offensive.

So we went on a trip to contact our old friends on their favorite *DarkWeb* forums.

In just a couple of days, we had already found most of them. Some of them were former members of the *LoU*

(Legion of Underground), with whom I personally had a certain affinity, part of my learning had been with them while polishing my skills. I can say I have never stopped considering them as superheroes.

LoU had a reputation for fighting for just causes, in 1987 they declared a cyber war on China, and although not the entire community was on their side, this type of action was nothing new for them.

Other acquaintances were also members of the *Chaos Computer Club, Cult of the Death Cow* and *L0pht*, the latter created one of the best tools to crack passwords that existed at the time, *L0phtcrack*, with this and John the Ripper I was a happy child.

Some of these friends had contact with more "activist" groups without fear of confrontation, such as *LulzSec* or *Anonymous*. However, no one offered their information openly... It was implied, though, that if we ever need their help, or if we wanted some of their members to participate with us, they would definitively volunteer.

Meeting these individuals brought back many memories, eagerly reading the *e-zines* created by hackers for hackers, *Phrack* and *2600*, and the obligatory discussion of the *Rainbow Series books*, published by the US Department of Defense *(DoD)* of the United States.

Some of them had met in person at *DEFCON*, the largest annual hacker conference. And I had any doubts that a couple of them had participated in "Operation Payback" in 2010, in defense of *WikiLeaks* with the mission "*Avenge Assange*", making one of the most controversial *DDoS* attacks in history, taking banks and credit card companies out of service. It was just brutal.

WikiLeaks' name was carved in history after publishing US diplomatic documents with ripple effects around the world. They began with video leaks in war zones of the US Army shooting defenseless people, including civilians, as well as documents that revealed the "collateral" damage caused.

Publications of documents, photos, and videos of the torture of prisoners during the Iraq war had a profound impact on American politics forever, but the *Cablegate* publications were the launch to stardom of *Wikileaks*.

More than two hundred and fifty thousand documents between the State Department and embassies around the world were revealed, one of the largest leaks in history. The information was read thousands of times, as everyone wants to know what Americans say about them at their backs.

For a nation that has been at war, invades and occupies territories constantly, and has done all this for nearly a century without rest, this number of documents may not even be surprising.

The government tried to use its power to get *WikiLeaks* out of circulation by putting pressure on all the companies that provided their services in one way or another, hence the "Operation Payback" was born.

This attack was special, it was not mainly carried out from *botnets*, but it was collaborative, even many people without computer expertise participated.

Tools created by real hackers were made available from the web or from an application. *LOIC (Low Orbit Ion Canon)* was one of the most famous, and it was as simple as *point & click* and *voila*, that was enough for any person to participate in one of the financial and technology industry decommissioning revenge attacks on *WikiLeaks*.

Companies like *VISA*, *MasterCard* and *PayPal* were hit, their websites went down for hours. In the cases of the credit companies, their main credit authorization operations were not affected, but those secondary processes, including plastic use authorizations that required additional validation, were. But Amazon was not affected, since 2006 it had been playing with its Elastic Computer Cloud, a fancy name for the use of self-managing virtual machines on demand.

But the name was not just marketing, they spent years creating an "elastic" infrastructure designed to grow automatically according to its demand for resources.

Jeff Bezos had given the green light to a strategy that would change the core of *Amazon's* business model and would mean one of the most profitable services for the company, its incomparable *AWS*.

This service, with its redundant and elastic capacity, was able to contain the attacks of *Anonymous*, and was the only company that did not dim the lights in the middle of the war, which meant an important test and achievement for its service.

Anonymous, a group that does not identify itself as a hacker group but as a "living online conscience," literally exclaimed that it didn't have enough firepower to affect the tech giant.

The characteristics of *Amazon's* philosophy over time became better, ten years later they again demonstrated the reason why their service competes in another galaxy, when in 2020 they received a denial-of-service attack for nothing more and nothing less than *2 to 3 terabit-per-second (Tbps)!* About 2.5 million megabits per second!

It is an incredible fact that the largest *DDoS* attack carried out in the history of the Internet against a company went almost unnoticed. Amazon did not even flinch, and only a few reporters found out and talked about it in the news.

But whether our acquaintances have participated in the Payback operation and in one or another social cause, they were characterized by their fair actions; but living under the ideal that *when tyranny is law, revolution is order*.

However, some friends of our friends had been caught ratting out their acquaintances to the *FBI*, due to the blackmail and retaliation they were subjected to, so you had to be careful and watch out on two fronts, even though it was something happening on rare occasions.

After a while, the small talk began remembering the good all times and anecdotes that could serve for several films.

The discussion heated up as soon as *Stuxnet* was mentioned.

8

DOWN HERE I'M GOD

"The Ferrari of malware," Warlock pointed out.

"My respects to its authors," said Xpl0it, reaffirming the complex and beautifully programmed Stuxnet code.

"By the way, this is Xpl0it. A wizard we've welcomed into the group," Warlock continued.

"I only recently met several of them, but I feel like I've known others all my life. A pleasure," Xpl0it indicated.

We didn't pay much attention to it, thinking that it was some allusion to the manifesto or he was trying to sound friendly, although what it really seemed to me was lame.

After a few minutes, however, we understood the reason. Xpl0it sent Robert and me a private message, informing us that he was Khalil, and that he saw our old pseudonyms. An interesting surprise after years without hearing from him. He was a ghost that showed up on special occasions but that we had not yet managed to identify.

After the classic greetings and hearing several times the most accepted lie in the world: "Everything is going very well", we continue chatting animatedly.

Everyone was fascinated with *Stuxnet*, perhaps if someone had participated in the creation of it, they would not praise it so much, but the talent that existed in the North American national security agency, *NSA*, was recognized.

Nobody got tired of talking about the malicious program and it really seemed like something out of a science fiction movie. It was the first known cyberweapon that had thrown Iran's nuclear program all over the floor!

Stuxnet, part of a family of highly specialized and advanced malware designed with a heavy focus on the Middle East, shares some code with high caliber malware like *Flame*, *Duqu* and *Gauss*, but *Stuxnet's* task seemed beyond fiction.

Imagine having a *malware* in 2009 with the capacity to penetrate a nuclear power plant that is not connected to the Internet in the Natanz region in Iran. It is presumed that some of the companies responsible for supplying the manufacturing and industrial components necessary for the nuclear plant were used as a route of infection.

The malware, after detecting that it had reached its target, it proceeded to stealthily infect computers, which had high security standards, taking advantage of not one, but four unknown weaknesses with *0-day exploits*, and proceeding to install a rootkit at the *kernel* level, with the ease that allowed it to do so by using trusted digital certificates of the operating system, which were usurped from *Realtek's* semiconductor company, without them having the slightest idea until the malware was disclosed.

And then advance in the network until it reached and affected the Siemens SCADA system, responsible for the industrial control program, and in charge of the plant's uranium enrichment. Proceeding thus with the manipulation of the centrifugal valves, increasing their speed and pressure to the point of destruction, while at the same time, interfering with the monitoring system to make the operators believe that everything was going great

A true piece of art.

A digital attack that impacted the real world, breaking down barriers forever.

But as wonderful as the *NSA* teams are and as many resources as they have at their disposal, the interaction between their brilliant staff is what gives them a real competitive advantage, and keeping our distance, that's what we were looking for with our old friends.

After paying homage to the "Ferrari" of malware, we got down to business. Robert, using his Internet alias *Kaffein*, recounted what had happened, and asked for everyone's support to, in the best of cases, identify or capture those who belonged to this gang. And in the worst case, at least get some of his pride back.

Knowledge was not the only thing we could share, the tools at our disposal and all the controlled computers around the world gave us some advantages. This gave us a different strategic position, and as expected, we are stronger together.

As Robert explained, RadeC0m noted:

"It is understood that an APT group is related to this type of attacks, if the addresses you saw it's relevant, and you got too far, we already need to be thinking how much they already know about us already."

"*Fuck*, we had a suspicion, about an elite hacker group, but where are they from? Russia, Iran, Korea, China?" Robert asked.

"One of them, or all of them, *LOL*," was RadeC0m's answer.

"You are well informed," Xploit said.

"We are traffickers of information dear, data is the new oil," RadecOm continued.

Robert just laughed:

"LOL."

Within these beautiful families of APT groups, different units with different objectives had been constituted, from espionage units to sabotage of private companies and governments. Each unit had its story.

An interesting and very public story occurred in 2014, when the group identified as APT37 settled accounts against *Sony Pictures*, for their rejection of the movie *"The Interview"* in which the North Korean leader was assassinated in a comedy.

Yeah, North Koreans don't have much of a sense of humor either.

For better or worse, they take their business very seriously, and the impact on Sony Pictures was brutal. The team extracted personal data from employees and executives, as well as copies of previously unreleased movies, including the famous Fury, starring Brad Pitt, which was viewed online at least one million times illegally.

They also published *Sony*'s marketing strategies and most of their employee salaries, causing public opinion to accuse them of showing little fairness and being discriminatory as it was demonstrated that the staff of their main executives earned more than six figures a year and that most of them were white men.

The attack did not end there as it continued with the destruction of computers.

They claimed to have obtained more than a hundred terabytes of data from *Sony*, although it could really be half, in any case, it was enough to damage the image and profitability of the company. It was one of the largest cyberattacks ever carried out on an American company.

Thousands of files were published exposing the names of the employees, their usernames, passwords, and digital certificates with their access keys. Salaries, bonuses, performance reviews, and even health records to name a few were also leaked. In other words, they took all their "masks" aways.

And to top it off, they extorted them for a sum of money with the promise of stopping the attack, but they published films prior to their release anyway, along with the revelation of sensitive emails with private conversations of the producers. It is now known that one of the executives mocked at how untalented Angelina Jolie was.

"Nothing's impossible, but it's easier to turn off the internet than to win a fight with guys like these," SysOP pointed out.

"It's easier and I don't need you to do it," Warlock joked.

"Sure, I remember your playful times with *ROOT DNS* and *Border Gateway Protocols*," answered SysOP.

"Shhhhhhhh, or I will be commenting on your entertainment with some satellites," exclaimed Warlock.

After the exchange of egos, we got back on track, and agreed to share the logs and data from the *DDoS* attack we had suffered, the one that affected Robert's entire residential area.

We would search for patterns on the command-and-control zombie computers, and everything that would lead us to identify the main perpetrators.

"Let's do an *RSA*," Anarchy said.

"LOL," Kaffeine replied.

"But with violence and noise," he continued.

Anarchy made a mocking reference to an attack on a major security company, known as *RSA Security*, not because of how they were attacked, but because of what the intrusion represented for the attackers.

RSA, famous for its *SecurID* physical token device, was responsible for generating a dynamic key every thirty to sixty seconds. It has been one of the most influential companies in the security industry, and owes its name to its three founders , who revolutionized cryptography.

RSA's encryption tools were licensed throughout the tech industry and ended up being used all over the world through any software imaginable.

Despite its tremendous history, after the change of main executive board, it is said that the *NSA* managed to influence them to "recommend" the adoption of a cryptographic system by default in their products, and with which they received millionaire contributions that many did not see with good intentions.

The problem arose when the industry detected weaknesses in this adoption recommended by the *NSA*, which would allow anyone with the "correct key" to have access to all the information that was supposedly protected.

They were accused of having created a «*Backdoor*» in cryptographic products through the *Dual Elliptic Curve* algorithm, which the *RSA* distributed through the *Bsafe* software to «protect» the security of computers, data, products and services for people and companies.

With a well-written marketing strategy, RSA adopted and promoted the algorithm in record time, and the *NSA* lobbied for the blessing of the respected National Institute of Standards and Technology (*NIST*), so it could be used in government institutions.

And with *RSA*'s influence in the tech industry, its adoption was accepted unsurprisingly, including by firewalls and virtual private network (*VPN*) providers, without raising any alarms.

It is not the first time, nor will it be the last that the *NSA* tries to weaken the available cryptographic systems, as the president of RSA himself, James Bidzos, indicated in 1994:

"For almost 10 years, I have come face to face with these people at Fort Meade. The success of this company is the worst thing that can happen to them. For them, we are the real enemy, we are the real target. We have the system they fear the most."

Bidzos was not wrong, the *NSA* has an entire program dedicated to this task called «*Bullrun*», and it is one of the most expensive, its figures reached around two hundred and fifty million dollars in 2013 alone, but nothing compared with the budget of its «*Consolidated Cryptologic Program (CCP)*», which represents 2.3 billion dollars investment.

The NSA has looked for ways to "incentivize" technology security vendors to provide a master encryption key or backdoor, in some cases hiding it through recommendations to adopt standards that are convenient for them.

They can achieve their goals by exercising the force of law, but this poses certain risks to the agency. In most cases it is easier for them to exploit the weaknesses of the private company's products. It has long been shown to have an unprecedented success rate without requiring any help from the manufacturer to achieve its goals; but data encryption has always been a sensitive issue for them.

In 2005, the *NSA* infiltrated the *Secure Sockets Layer (SSL)* and private network (*VPN*) protocols without difficulty. Years later, in 2013, *Google* replaced the *SSL* certificates from 1024 bits to 2048, because it no longer considered them secure, after seeing the effort that the *NSA* was making so that no encryption was beyond its reach.

For this agency, where nothing is impossible and secrecy is its way of breathing, cryptography is its greatest challenge and enemy, and to counteract it, they focused a large part of their effort, which allowed a glimpse of the head of that cautious monster.

But that *RSA*, despite its fame, relationships, and high security standards, suffered an attack worthy of a Hollywood movie. The most surprising part of the attack that Anarchy referred to was not the violation of the security of a company characterized by being a benchmark in protection mechanisms, which had brilliant minds and *state-of-the-art* technology for the perimeter and their workstations. The surprising part wasn't either that services worth forty million dollars in *SecurID's token* were affected, nor that it ended up impacting the company for a total of more than sixty million dollars. The reason behind the attack is much darker.

The attack began with a well-known technique, a well-crafted *spear phishing* targeting select employees, and continued with a *0-day exploit* inside an Excel file, which downloaded *PisonIvy* and were used it as its access tool *(RAT)*. After gaining total control of the computer, they escalated privileges until reaching the family jewels.

And here we are not talking about *Bsafe*, but about its distinctive brand and highly protected intellectual property that generates dynamic codes for *SecurID*, used to protect the most sensitive data and access by thousands of companies around the world.

Navigating within the network to achieve this goal is remarkable, considering how it went unnoticed among the security systems of a company like *RSA*, with the biggest and most developed mechanisms of the technology industry.

What reduces the impact and magnitude of this incident is to know that this was just one of the steps to go after the real target: *Lockheed Martin!*

Nothing more and nothing less than one of the main suppliers of the United States Navy, builder of missiles, and F-35 fighter planes, and the "little" *hack* to RSA represented a step to obtain the holy grail of the keys of the *SecurID tokens*, and thus violate *Lockheed's* remote access (*VPN*).

Using the RSA as a queen sacrifice to kill the king? This was something amazing.

Lockheed Martin claimed to have detected the remote intrusion almost immediately, and that no significant damage had been suffered, which sounded more like a political message as they proceeded to disable all access for over a week.

This was a reminder that no one, not even the industry benchmark, is free from being hacked. It's just almost inevitable to lose a fight when the very systems that are meant to protect you serve as attack bridges.

Security is built on trust, and when trust is lost, there is no security.

Due to the indirect nature of the attack, Anarchy linked it to our next steps.

After sharing some information, we agreed to see each other in several days, so we proceeded to say goodbye and hope to bring good news.

The wait did not last long, but the result surprised more than one.

9

IGNORANCE IS BLISS, UNTIL IT'S NOT

O ne of their command and control centers is in your city, dear Kaffein," my old country, Khalil, said as we greeted.

"Nooooooo kidding," responded Kaffein.

The geolocation indicated it was in the center of the city. A medium-sized technology services company, from where they sent instructions to thousands of bots for *DDoS* attacks. Due to its operations, it wasn't just another botnet. Some more relevant things were happening there.

"Does anyone have a friend at *The Equation Group* to lend us a few toys and save us the trouble?" I said jokingly.

They all laughed.

"*The Shadow Brokers* then?" I continued.

Nobody laughed.

The power of the *NSA* within the computer world is comparable to a god, under the motto *"Getting the Un-gettable"*, its *TAO* (Tailored Access Operations) unit, to which *The Equation Group* belongs, seems to have no limits, but each god has his devil.

And to the NSA, the devil is called *The Shadow Brokers*, an organization that gained its name and its position as the NSA's favorite public archenemy, revealing dozens of their tools, methods, and exploits. Worse still, they are mercenaries who sell them to the highest bidder, offering exploits that could allow them to enter any system in the world.

The NSA catalog of products for the intelligence agencies is set with prices and models, just like department store item, but it is exclusive to the VIP group of super friends known as the Five Eyes, the intelligence network formed by Australia, Canada, New Zealand, the United Kingdom, and the United States.

The Shadow Brokers also had their catalog of exploits, from the NSA itself, but this was not limited to members of the Five Eyes, but to any individual willing to pay the price.

Sometimes reality writes the best scripts.

Interestingly enough, this catalog held a public auction before putting it up for sale inviting the NSA to keep its toys secret, of course, as long as its bid was the highest. They claimed that they were not irresponsible criminals and that they were aware of the damage this would cause, so it gave the NSA an opportunity to stop the leak of their tools for the right price.

They received no official response from the NSA, so they went about their business, changing the auction model to an *à la carte*, and later to a subscription model to receive new *exploits* on a frequent basis.

They acted like good entrepreneurs who listen to the market and improve their services, only to then deliver the final blow by simply releasing the exploit kit publicly.

Those releases, known as *DoublePulsar* and *EternalBlue,* were the most damaging for the world, as they were later used to attack hundreds of thousands of computers.

The Shadow Brokers were also responsible for publishing *NSA* operations against nuclear plants in Russia, China, Iran, and North Korea, as well as bank intrusions in countries such as Qatar, the United Arab Emirates, Palestine, and others. region of.

One of the most surprising, published actions was the revelation that they had attacked the heart of the entire *SWIFT* network through the *SWIFT Service Bureau,* in a way that allowed them to have a view of almost all international transfers in the global financial system.

But the action performed by The Shadow Brokers is one of the most incomprehensible in the entire hacker community. A team capable of embarrassing the largest espionage agency, looting its precious material, and weakening its power of action, or at least delaying it for a considerable time, dedicated itself to revealing and exposing these tools instead of taking advantage of them in a silent way, knowing that they will suffer retaliation by the NSA for as long as they existed.

Selling that secret weaponry is the same as selling the goose that lays the golden eggs, unless their intention has always been to expose the tactics, techniques, and procedures of the *NSA* while using economic excuses for their real purpose.

It is interesting that an agency known for its mystery and secrecy has been compromised in this way, and no longer has the privacy and anonymity that they would like, since some intelligence centers are well known, such as the case of *TITANPOINTE* at *33 Thomas Street* in New York City, a building 167 meters high, without windows and with enough food for fifteen hundred people for two weeks, and capable of withstanding a nuclear attack, or the well-known room *641A* in San Francisco, operated by AT&T to facilitate the surveillance program for the *NSA*, given its strategic location to capture Internet traffic.

There is no way to escape from these agencies if they want to know about you, basically consider all your data, network and digital equipment's compromised. Luckily, the *NSA* has more important things to attend to, but their claws or back doors are in every unimaginable component, from keyboards, hard drives, network equipment, cell phones, everything. Even on the very software and hardware that is supposed to protect us, such as *routers*, *antivirus*, and *firewalls*.

That makes the act carried out by The Shadow Brokers more incomprehensible. They had access to these tools and, thus, a special power in the virtual and physical world, in addition to being imperceptible and with the ability to violate everything around them; something like that reminded us of the attack carried out by the Russians with "The Thing".

American intelligence was not prepared for this attack, "The Thing", it was as enigmatic as it was surprising, from the year forty-six to fifty-two, all the classified conversations in the house of the ambassador in Russia were listened to through this device.

This attack was unlike any other espionage mechanism and did not use electronic components. Therefore, even if Faraday cages were used to protect against *TEMPEST* attacks, and thus prevent espionage through electromagnetic waves, it would have been useless.

It didn't even use cables or batteries, and they discovered it by chance. A carefully hidden device that works like a retroreflector radio to extract the sounds in the environment.

The espionage was a work of art, it began through a social engineering attack with a donation from the Soviet school of "The Great Seal" to the United States, inside this there was a trojan that enabled the "back door" to listen to conversations for years.

Inside "The Great Seal," there was this surveillance beauty that captured sound waves through its resonant cavity, which were then sent out when stimulated by a radio frequency that the Russians broadcast from a van. How about that? Point for Russia!

These were tools that we could only dream of, so, coming back to our reality, we had concluded that keeping track of the attackers had had unintended consequences, and that we should respect the enemy's "*Know-How*", limiting our exposure in the net.

"So guys, so much partying and playing vigilante and one or two other bad steps out there, and nobody has a 0-day that they want to share," Xpl0it exclaimed.

"Let's not get stingy now. We need to understand what happens in that technology services company from where they sent the instructions to the bots for the DDoS attacks," I seconded.

There was a strange silence, it was not only to use the best exploit, but also to consider the best strategy considering the level of the enemy. After the pause Kaffeine said:

I have an idea, let's put the electronic equipment to sing."

"I'll tell you later."

"WTF?!" everyone responded almost in unison.

10

THE WORLD AS PLAYING FIELD

While we waited for Robert to feel like sharing his idea, we turned to that dark part of the Internet where all kinds of things can be found, the infamous Dark Web.

There, it is possible to find services and exploits for sale that can harm systems considered safe. Products, let`s say, very similar to the offers made by *The Shadow Brokers*.

Ukraine experienced a cyberattack in 2017 that used one of the tools made available by the NSA's archenemy, called EternalBlue, which exploits a vulnerability in Windows Server Message Block (SMB).

That *exploit* allowed the creation of the *Petya* and *WannaCry* malware, and between them and their variants, they managed to infect more than sixty-five countries and caused billions of dollars in losses.

WannaCry and *Petya* behaved similarly yet differently at the same time, the former acted as a true ransomware encoding data and requesting a ransom payment, and the latter was more of a data and system destroyer without much interest in recovery. Both shared their tactics to achieve their propagation and success.

A modified version of Petya was the one that attacked Ukraine in 2017, suffering one of its worst cyberattacks, even considering the attacks on its electrical system in previous years. This attack spread at an impressive speed, it took less than a minute to kill the companies' network, and to say goodbye to your systems and data.

At that rate, it devoured more than twenty-two banks and hundreds of companies, for the second largest bank in Ukraine this meant losing 90% of its computers.

It is estimated that more than 80% of companies suffered the attack with irreparable damage to their systems, categorizing it as the largest cyberattack on a nation to date.

It was the equivalent of a digital atomic bomb, needing only a few seconds to render inoperative the financial system, including its *ATMs* that displayed the bogus ransom message.

They also lost the ability to process payments in stores, and the transportation and purchasing system in general was impacted, the clinics had to return to pencil and paper, the airport information system disappeared, as well as the electrical system, in short, nothing was safe. It was like The Flintstones comics.

It was a severe blow that put multinational companies such as *Maersk, FedEx,* and *Merck* in trouble, impacting their services worldwide.

In the case of *Maersk*, the marine transport giant and the world's largest container operator, they estimated direct losses from the incident of around 300 million dollars, and it took them more than three months to repair four thousand servers, including forty-five thousand personal computers, and two hundred and fifty applications.

For one of the leaders in air freight, *FedEx*, the cost was even higher, with $400 million in losses and six months for a full recovery.

But the injury to the pharmaceutical company Merck took the crown, with 670 million dollars and the loss of fifteen thousand computers, which caused the production of essential vaccines to stop on the production line.

Although this attack was not planned to impact beyond Ukraine, the interconnection with suppliers, supply chains, and international presence allowed for an impact beyond the borders.

Individuals and companies are less and less able to escape the collateral damage of cyber wars. Unlike the territorial demarcations on the geopolitical map of a conventional war, these lines become very blurred with the wonder we call the Internet.

The malware entry vector used in Ukraine took advantage of a backdoor previously implanted in the *MeDoc* accounting system update.

The *MeDoc* software basically used by the whole country for tax payment allowed to have a secure entry point to almost all of Ukraine.

After the malware gained access to a computer, it would take advantage of *Mimikatz* to extract passwords from memory to access other computers on the network.

In this way, if a computer was protected because it had the patch for *EternalBlue*, it was later compromised by the theft and reuse of credentials on other computers.

A vulnerable computer and you can say goodbye to your network.

By mid-2018, another little gift awaited Ukraine, this would be another year for the country to be used as a playing field and practice, but this time they were lucky.

More than five hundred thousand routers were infected with the malicious *VPNFilter* program, mainly in Ukraine, programmed with a *"kill switch"* to disconnect them from the Internet. Luckily for them, *Cisco's* security division spotted the problem, and alerted major *router* companies about it.

While for *NotPetya*, as they called the variant used in the Ukraine, it was too late, the systems of hospitals, power plants, airports, banks, gas stations, electricity companies, payment systems were already collapsed.

WannaCry, also a son of *EternalBlue*, also infected more than two hundred thousand computers in more than one hundred and fifty countries.

In England alone, sixteen hospitals had to cancel all non-urgent services, as well as twenty thousand appointments, and six hundred surgeries had to be rescheduled.

The damage could have been worse and created greater chaos in the world, but thanks to Marcus Hutchins, a security researcher who was analyzing the malware, noticed that it was trying to communicate with a web address that was not registered, and when proceeding with its registration discovered a "kill switch" in the code helped stop its spread.

In the end, one vulnerability, two malicious programs, billions of dollars in damage.

NotPetya alone is credited with more than ten billion dollars in losses, making it the costliest and most destructive cyberattack on record.

These digital attacks that have had a greater impact in Ukraine, suffering from distributed denial-of-service attacks, electrical blackouts, destruction of equipment and damage to services and infrastructure, point to the *APT SandWorm* group as the main culprit, with ties to Russia.

Technological evidence, which has not been easy to obtain and has required the systematic evaluation of various attacks, comparing similar tactics, techniques, and procedures, points one way or another to this APT group, despite the false flags that have been left intentionally intended to mislead forensic investigations

We had no intention of doing harm with tools of this caliber in our possession. We knew that with these available exploits you could do many things, not necessarily destructive. In our case, capturing some of those responsible for a bank robbery in Chile was a very exciting task, but it became personal a while ago.

We were open to any arsenal of weapons to catch them that we found on the Dark Web.

"Well, then, as we set up our equipment for this, are we open to everything?" Radec0m asked.

"We have a preference for physical exploits over network ones this time, but we may need a bit of everything," I replied.

"Target: new NSA catalog," Ra-dec0m joked.

"Yeah, right," was SysOP response.

11

EVERYTHING BEGINS WITH CHOICE

We wanted to be outside the network as much as possible, we already knew the level of the enemy against whom we were fighting, and, even if the technology services company was just one more victim, we would not lower our guard; this enemy was very dangerous.

We explored off-the-shelf physical tools to obtain data from the target, be it placing traps in the network, screens, keyboards, etc., but all of them were perceptible to a good connoisseur. These were no match for the *NSA* ghosting tools.

One of those tools that caught our attention looks like a normal *USB cable plug,* and it can be implanted in any device such as a keyboard or mouse. This powerful tool called *COTTONMOUTH* has a radio frequency transceiver that allows you to create a data bridge miles away.

It is known that at least since 2008, the *NSA* used it as a secret channel of radio waves for the transmission of data, allowing them to have access to those computers that sought protection because they were not on the Internet. A false perception of security against enemies of this caliber.

"I got a copy of *RAGEMASTER*, but I have my doubts," Anarchy pointed out.

"Share the photos and features," Xpl0it replied.

"It doesn't work, it's a thousand times bigger than the original model," Kaffeine pointed out immediately.

"I still haven't got the TAO shop to meet your needs, my lord," Anarchy answered uncomfortably.

"LOL, calm down guys, let's keep looking without complicating ourselves," I commented hoping to calm the spirits and ease the bad mood.

Our interest was also piqued by *RAGEMASTER*, a tiny RF retroreflector that can be hidden in a standard computer video cable and can remotely replicate anything the monitor presents.

Despite the challenge of emulating these devices, we had vast information and material to improve the copies we found for sale, although it required a physical visit to the facilities, and we did not count on the *CIA* to do the dirty work of the *NSA*. This was another challenge to overcome, but we would deal with one problem at a time.

We talked about various ways of putting our ears to what was happening in that company without exposing ourselves and, depending on the conditions, it could be relatively simple. So the conversation focused on side channel attacks in search of taking advantage of the electrical emissions of the equipment by any means.

"If we want to limit network intrusion to a minimum, we should think about *TEMPEST*-type tools," Anarchy pointed out.

"Finding something of this high quality on the *Dark Web* is like trying to find a plutonium dealer. It seems very complicated to me," Radec0m answered.

"But this is the way, to spy through the leak of emanations, be they electrical, sounds, vibrations, waves, *whatever*. I already had a bad time with these guys to expose ourselves again. I will be sharing what I promised, this is the way," Kaffeine answered.

"*¿What about PITA*? I understand that the Israelis have managed to create in a budget a device capable of extracting cryptographic keys by capturing electrical emissions," Warlock asked.

"Yes, but it is very experimental, and you almost have to be next to the computer, it would not be useful in this case. I have my contacts at Tel Aviv University," Anarchy pointed out.

"*BITWHISPER?*" he asked again.

"It would not be an option either, it physically requires the installation of the malware before taking advantage of its capabilities to capture data through the heat emissions of the equipment. Besides, you should also be very close, unless someone else knows something that I don't." added SysOP.

"That's super cool! Using heat patterns so the other equipment to interpret the data flow. That is fucking genius! Well, just adding that it really ends up being a covert channel attack as it allows sending data in addition to extracting," Xpl0it added.

"If we consider *AirHopper*, data can be captured and sent via the cell phone's radio receiver, and this allows it to be used over greater distances, up to a few meters," Xpl0it continued.

"Well, focus, there are very cool attacks that are not within our reach or are not useful in this scenario, including *AirHopper* due to its limited data capacity, let's explore anything else we can make work out," I stated.

"But that's where the shots go," Kaffeine added.

"We need those features, but make it possible at a greater distance," he continued.

These options would allow us to remain virtually disconnected from the victim, so it is interesting to explore how to infiltrate and extract data by taking advantage of the electromagnetic emissions of the equipment.

But we always found disadvantages in terms of their capabilities with respect to distance and use, or doubts about the reliability to perform a successful attack, and more importantly, it could not be executed fully remotely effectively without at least a first physical visit, and this prevented us from obtaining the secrecy we desired.

"I think we should do two things. First, put ears inside the place, and then analyze the effects of the digital components on the physical ones, so that we can make them sing, as I told you before," Kaffeine said.

Robert had the peculiarity of having fun explaining things in a complex way, just because he felt like it, like saying the scientific name of a plant instead of calling it by its common name. It was the kind of allusion that could bore a person who didn't know as much as he did, or who wasn't that patient.

But what he said was never nonsense. He had the ability to dismember very complex topics and take them to the most plain and simple language, like anyone who has a great command of a subject.

"Feet to the ground, stop joking." SysOP fired.

"Ok, I've been analyzing the situation and I think the first thing we should do is find a way to listen inside the place. Does anyone ring a bell the *visual microphone*?" Kaffeine pointed out.

"Yes, that can work, and it is possible to use it with a good distance," Xpl0it replied.

"Are we going to start building one?" Anarchy asked.

'Something better, a *Lamphone*," he replied.

The "*visual microphone*" is an interesting method of capturing sounds via video.

The waves produced by our voice generate unique characteristics in everything they touch, and if we were able to identify those small variables, we could hear whatever was said in a room from a safe distance; It sounds like science fiction, but it's very real.

Imperceptible to the human eye, a high-speed camera detects small vibrations on the surface of an object, and through special *software* the sounds of the place can be reproduced.

Some *geeks* at MIT published a proof of concept showing these capabilities a few years ago, and it's only gotten better from there. To the point that a version called *Lamphone Attack* is capable of spying with the use of a laser microphone through electric bulbs tens of meters away.

"We're going shopping," Warlock exclaimed.

"Hold on, hold on, hold on, Kaffeine…. And the story of making the electronic components scream?" I asked for.

"Ahhhh my dear Watson… here we will be inspired by "*The Thing*," he replied.

"Wow! I didn't know that we are capable of creating such an artifact," I replied.

"I did not say that we would create "*The Thing*", but that we would be inspired by it, that is, in its technique to extract information, using a secondary radio channel to transmit data, without using Wi-Fi, Bluetooth or any Wireless communication method," Kaffeine explained.

"And what are we talking about?" Warlock interrupted.

"Funtenna!" Xpl0it exclaimed, as if it were a guessing contest.

"A Funtenna attack! That means putting the electronic components to sing," he continued.

"Indeed, my friend, but first things first. Later I'll tell you how I plan to do so that we can use it while maintaining the protection and anonymity that we want," Kaffeine replied.

"Let's get to work then!" SysOP yelled excitedly, as we were now moving from planning to action. So, we went to get the necessary parts to build our *"lamphone"*.

Our super spy tool was ready in a few days, and we took turns trying it out in harmless pranks with those friends who shared the island. It worked perfectly when the weather conditions were good while recording conversations through different types of light bulbs.

We were very excited to put the plan in motion for our first battle. We located a privileged position and we aimed at the only visible place of the technology services company. We were just steps away from knowing in depth about one of the heads behind the previous attacks that we had suffered.

But our lamphone did not throw anything, tests went back and forth, and we did not get results, we checked the configurations and components again and again, we tried other companies around it with success, but when we returned to our target, the result was negative.

What we had not calculated during our excitement was that the bulbs of our target were adorned with some rigid craftsmanship that prevented us from having a full view of the electric bulb, and that the windows had some type of laminated protection, preventing these microscopic movements from being captured.

"What the fuck, how the hell didn't we validate this before?" Warlock exclaimed.

Emotions made us neglect rigorous analysis and we wasted time on something that got us nowhere.

But the plan had not changed, we needed to carry out a physical attack without exposing ourselves, with equipment that we could discard without leaving any trace.

"Kaffeine, do you want to sing now?" Warlock asked.

"Loud and clear," he replied.

The Deus Ex Machina 2.0 operation had begun.

12

UNFINISHED BUSINESS

Since we wanted to do a very stealthy surveillance, while taking care of being undetected, our inspiration in *"The Thing"* would be the reference to carry out a remote attack and extract information without exposing ourselves to any monitoring system that might exist.

We really didn't know what levels of security we would find, but we no longer underestimated the enemy. We had learned our lesson.

Our attack was not as rustic as *"The Thing,"* but it was still magical. Our *Funtenna* would be in charge of modifying the electronic components of the circuits through simple lines of code so that they would behave as an antenna, and thus it'd transmit the data through radio waves, avoiding detection by the protection mechanisms generally used even in highly protected networks.

Nowadays nothing moves without *Silicon Valley* parts, but since these components weren't built to work that way, the signals didn't get very far, so just like the Russians had done by stationing themselves outside the American embassy, we had to get closer to implant the malware with the Funtenna attack, and also later to capture the information.

We rambled some ideas without reaching any agreement, until Pwnjuice commented:

"I have it."

"Let's see genius, what do you have? Did you get the laser to listen through the walls? so we can solve this? I hope it's better than our Lamphone," Anarchy joked.

"Available soon on your favorite Dark Web site, including noise and vibration suppression," Pwnjuice replied.

"LOL!"

And then he continued.

"While you guys have been talking nonstop and to no avail, I've had some private talks with Kaffeine, and we've checked outside cameras near the company," Pwnjuice added.

"We have seen that they have outsourced technical support tasks for computers and printers. This is how we're going to enter without exposing ourselves directly, this is our *RSA*," Pwnjuice concluded.

"This idea can work," I said, and the rest approved.

So, we got going.

It took no time at all to send a *spear phishing* to one of the underpaid technicians, telling him to update our target's printers with a new patch to improve their performance and security, the request was perfectly crafted, including the ticket number sequence assigned to the case.

The new "patch" had been compiled by Xpl0it and did not make any improvements to the printer *software*, other than displaying bogus update screens.

The rest of the team helped by providing operating system *exploits* that allowed our "patch" to run and ensure privilege escalation.

"Well... there we packed some little gifts for different manufacturers," Pwnjuice said.

When installed correctly, our malware altered the electrical properties of the printer's components, turning it into our personal radio transmitter.

In a matter of days, we had the printers emitting a copy of every printed document via radiofrequency without anyone being able to detect it.

It was our own version of the *DocuColor* patterns that are printed on each sheet without being visible to the human eye, and it was helping us track each printed document.

Our malicious program was also in charge of passively detecting the telephones that were via voice over IP and carrying out the same exploitation process now automatically and without interaction, in this case taking advantage of the *exploits* incorporated by the guys to the Cisco system. So, in addition to capturing telephone conversations, we were able to capture network data by performing a passive mode *sniffing* on the equipment, allowing us to obtain all the segment traffic.

We didn't want to be intrusive by attacking computers, so we limited our targets to those that don't usually have any oversight.

Now, with the equipment emitting information via radiofrequency, we needed to be close enough to capture the data, and that's another story.

Again, we didn't want to be so obvious, but Kaffeine was already working on this part of the plan, stating:

"Ok... the first idea is to dress Warlock as a plumber and have him infiltrate unclogging the bathrooms in the place," he joked

"LOL, clown," he replied tersely.

"Well, the second idea is less humble, even against our ideals, but we are at war," he said.

"Spit it out," Radec0m said.

"As you know, it's always a bit tricky to get close without exposing ourselves. So, we'll send some robots," he said.

"What the hell are you talking about?" RadecOm asked.

"We'll use drones to capture the airwaves," was the response.

"It would be necessary to get an arsenal to have them enough time in flight and capture enough data, how will we control that?" SysOP asked.

"Leave that to me," was Warlock's less than humble reply.

The idea was simple, to get multiple "borrowed" drones without the consent of their owners, so Robert visited a park frequented by young socialites where dozens of drones were often seen in the air at the same time.

The reason for obtaining the drones in this way and not buying them was to eliminate any traces that could be used to identify us.

Robert used a Skyjack drone that, among other things, uses the terrible Aircrack-ng, famous for subjecting any wireless network to its mercy.

With him in the park, and from a safe distance from his car, he took control of multiple of these flying devices without difficulty. When the confused owners realized that they had lost control over their drones, Robert was already far away.

Although we were not in favor of this type of "indefinite loans", this time the end justified the means. We knew we were not playing in the minor leagues.

Later, Warlock was in charge of programming them to fly sequentially, while one drone collected data, the rest recharged energy, an entire automated spy factory.

With the drones modified to capture the waves of our *funtenna* attack, we began to monitor all the activity for several days.

With printers, telephones, and network data at our mercy, we gradually understood the mechanics of the place.

I wanted one last touch before moving on to the analysis phase, just in case.

"I think we can continue outside the net, while we expand our capture capacity," I told them.

"What do you have in mind?" Kaffeine asked me.

"I can get a modified version based on *KeySweeper* for which the level of cryptography or brand of manufacturer used is irrelevant, and in this way, we can obtain all the data that goes through the wireless keyboards" I said.

"That would be nice, a wireless *keylogger*, but would this not mean exposing ourselves too much to install it?" Xpl0it replied.

"Nope. We'll use our support technician to replace the printers' power supply with our version with the keylogger," I concluded.

"Great! Let's get to work," he replied.

So we decided to expand our listening and use a more powerful version of *KeySweeper*. So, we asked our human *"Bot"* to change the *Power Cords* of the printers, based on a new, more "modern" model, with supposed improvements in their energy-saving capabilities.

Our dear human *«Bot»* never suspected anything, he simply limited himself to completing the tasks of his *Ticketdesk* as soon as possible, so that he could meet his performance indicators and not be penalized.

As soon as he changed the *Power Cords*, we began to capture most of the data from the company's wireless keyboards, now via the *GSM* cellular network.

Without touching a computer, we had them compromised to the core.

13

Death, Taxes, and Being Hacked

We had material to entertain ourselves with all the data captures we've subtracted.

"These assholes not only serve as intermediaries for dirty jobs as a command and control center for *DDoS* attacks, they are also implanting backdoors to those who hire them," Robert pointed out.

"Fucking irony," Warlock replied.

"It's a good cover-up, these guys serve dozens of companies, but they're probably only interested in a couple. We opened a pandora's box, but we still needed to identify if they are victims or perpetrators," SysOP said.

"Guys, something's up," Robert said.

And he left the chat.

Robert was probably the most paranoid of the group, even an *ICMP ECHO* between his computers was enough for him to get nervous and make him disappear. We didn't worry when he disappeared for several days.

Anyway, the rest of the team agreed to stop monitoring until further notice, we already had a good amount of work to do with the data to analyze.

Sometime later, we met again, and we still hadn't heard from Robert, nor from Xpl0it. But we did learn something new that excited us and made us worry at the same time.

The analysis of the data and the characteristics of the malware detected in the technology services company show shared similarities with a known group: *APT38*, also called *Bluenoroff*, located in Asia. Now we have a face for our enemy, a North Korean state-sponsored threat actor.

The similarities were mainly found in data destruction tools and encryption keys for communication.

This APT group, belonging to the unit with the artistic name *"Lazarus"*, shows an interesting nickname in a practically atheist country.

So, in addition to espionage and sabotage, a new weapon has been added to his arsenal: robbing banks and financial institutions.

And given the similarity in the tools of the attacks on the financial system, it would lead to the conclusion that they are responsible for more than 20 attacks on banks in different countries.

The identification of the group quickly confirmed our fears and their skills.

"Check out this news stream," Ra-deC0m said, sharing the video link.

Kaffeine was being pointed out by the authorities as the main hacker accused of stealing dozens of companies and being a participant in attacks on banks in several countries using the *SWIFT* network, according to the news.

Twelve million dollars from Banco del Austro in Ecuador in 2015, ten million from the Bank of Ukraine in 2016, sixty million from the Bank of Taiwan in 2017, and ten million from Banco de Chile in 2018, were some of the accusations.

"He's been framed!, APT38 made him look responsible for his attacks as possible retaliation in a way that the authorities could find him." Warlock exclaimed.

Everyone was cold and confused trying to understand the situation.

"And where the hell is Xpl0it?" SysOP asked.

"It's too much of a coincidence that he's gone just now," Warlock replied.

"He has always been a strange character with his ups and downs in our relationship," I added.

"Would he have helped frame him?" SysOP asked again.

"We can have a lot of differences and confrontations, but why would someone do something like that?" It doesn't make sense to me," I said.

"Envy and hate are often more powerful motivators than love," RadeC0m said.

After an awkward silence I replied:

"We are speculating, we don't know if he has been arrested either and they are simply preparing the charges."

What we did not know either was that we had been infiltrated, one of the first computers that Kaffeine had access to was a honeypot, which also had a tool similar to a "canary token beacon", to monitor anyone who had dared to arrive so far; and with the abilities of this *APT* group, we knew that even hiding behind multiple proxies or the same TOR network was not a guarantee of absolute security.

As Kaffeine's access to the first server had been made before taking the drastic measures, *APT38* already knew everything they needed about their target, and they patiently waited for the ideal moment.

APT38 was able to see what was happening carefully, and when they realized that we had managed to identify how they were taking advantage of the technology services company to install malware in other institutions, they decided to act.

It was an important piece of their arsenal that served them for different purposes, and its public disclosure could provoke an exhaustive review in all the providers of banks, telecommunications companies, and services around the world, having a significant impact on his espionage and theft capabilities.

So they chose to go back on the offensive with more force than ever, and make the message clear to whoever thinks they are capable of challenging them.

They proceeded by attacking some small banks, this time adjusting the codes and altering the connection logs so that it would identify Robert's computer as the point of origin.

Cleverly, his *false flag* to fool investigators included a record where the program used to alter and manipulate the *logs* had apparently tried to run unsuccessfully, thus avoiding successfully completing the evidence removal process.

The visible IP address would also not point directly to the final address, since such a fact would raise suspicions about the skills of the attackers.

So the researchers would see one of the *hop points* in the log files, where a review would find another "bug," in which an *IP* was not removed, thus providing Kaffeine's implanted IP address.

The same thing happened in another bank, in this one they did not replicate the same "errors", but they did leave a code that would allow them to be related to the previous attack, so that this one would not go under the radar. After that, they proceeded to destroy some important computers.

They then planted evidence at the *hop point* to link it to past attacks on financial institutions where millions of dollars had been stolen, which would lead to an indictment not only of these recent intrusions, but of all cases identified as using similar techniques and procedures.

Like their previous attacks, it had been almost perfectly planned and executed.

The recent attacks had received the attention and help of international intelligence departments concerned about the growth of this threat around the world. Even Santiago, the friend who involved Robert in the investigation of the attacks in Chile, was an active participant in the investigation.

Some investigations take years to carry out, and some reached little progress, in this case, given the ease of the evidence left by APT38 and the quality of the team behind the investigation, in a matter of days they had an entire file built.

The intelligence institutions boasted of the success of the operation, which was only possible through their technicians and international cooperation, resulting in the capture of those primarily responsible for the attacks on the *SWIFT* system. "Banks can relax," they were commenting on the media.

" They have updated the defendant list and are going to formally prosecute the hackers, including Kaffeine" RadeC0m exclaimed.

While presenting the alleged perpetrators, RadeC0m continued:

"Who the hell are these? Are the rest of us still here?" he asked concerned.

It was a complicated situation, Kaffeine arrested, Xpl0it missing, and the rest wondering: "who will be next".

It wasn't until I took the courage to watch the news that I was able to have a better understanding. It took me a while to explain to the group, my fingers and legs were shaking uncontrollably, I felt dizzy, and I could barely write.

When I managed to regain strength, and I finally could explain to the guys:

"That......isn't Kaffeine."

"WHAT! What are you saying? How come he is not? What is happening?" They all seemed to say at once.

"I've known him since he was very young, that's not him." They don't even look alike. They must have identified his pseudonym online, but they haven't found him. Kaffeine must have hidden his IP address well, or as a last resort he will have diverted attention elsewhere.

"And I bet the new group of accused are scapegoats from the technology company."

"And Xpl0it? Is he there?" You know him too, don't you?" SysOP asked.

"Yes, I know him. He is not in the group of individuals presented, but neither is his pseudonym part of the accusation," I answered.

"Strange... very strange..." SysOP replied. I had the same thought, but I did not voice it. My mind had trouble finding positive reasons.

What mattered the most was that we knew that Robert was fine, or so we hoped, he was able to identify the danger before it was too late and managed to get off the train tracks before it ran over him.

Our fight would not stop, what had happened intensified our desire, it would no longer be those casual friendships, but a working group with a mission.

RED PILL

A new *"Shadow Brokers,"* united by fury, duty, and ideas; and those don't stop with prison or death.

As adversity has always allowed forging better men and developing better ideas, in this case, it had forged a group of brothers to fight for a common vision.

"We have lost this battle..., the errors of the past came to hunt us, and that is inexcusable against elite hackers, but this is the beginning," Warlock said.

And as if Robert had been listening to us the whole time, we received a message:

"This is our world now... the world of the electron and the switch... the beauty of the baud..."

"Yes, I am a criminal. My crime is that of curiosity. My crime is that of judging people by what they say and think, not what they look. My crime is that of outsmarting you, something that you will never forgive me for. "

"You may stop this individual, but you can't stop us all... after all, we're all alike."

EOF

ACKNOWLEDGEMENTS

To my wife, Anabel, because she puts up with my disconnection from the present day and night to lose myself in thoughts and reading. She has supported me with her patience, love, and advice. To my first child, Olivia, who will be born around the time this book is published, and it fills me with joy to know that I will soon have her in my arms. To me, my parents, Altagracia, and Carlos, who have given their best so that their children become the best we can be. They are my sources of inspiration, integrity, love, and greatness. My siblings and family, who fill me with joy in every encounter, great and unique beings, who *can count on me, not up to two or up to ten, but count on me,* as   Benedetti said in his poem.